THE SCIENCE OF SLEEP

KHUSHDIL MIR

Made with ♥ on the Notion Press Platform
www.notionpress.com

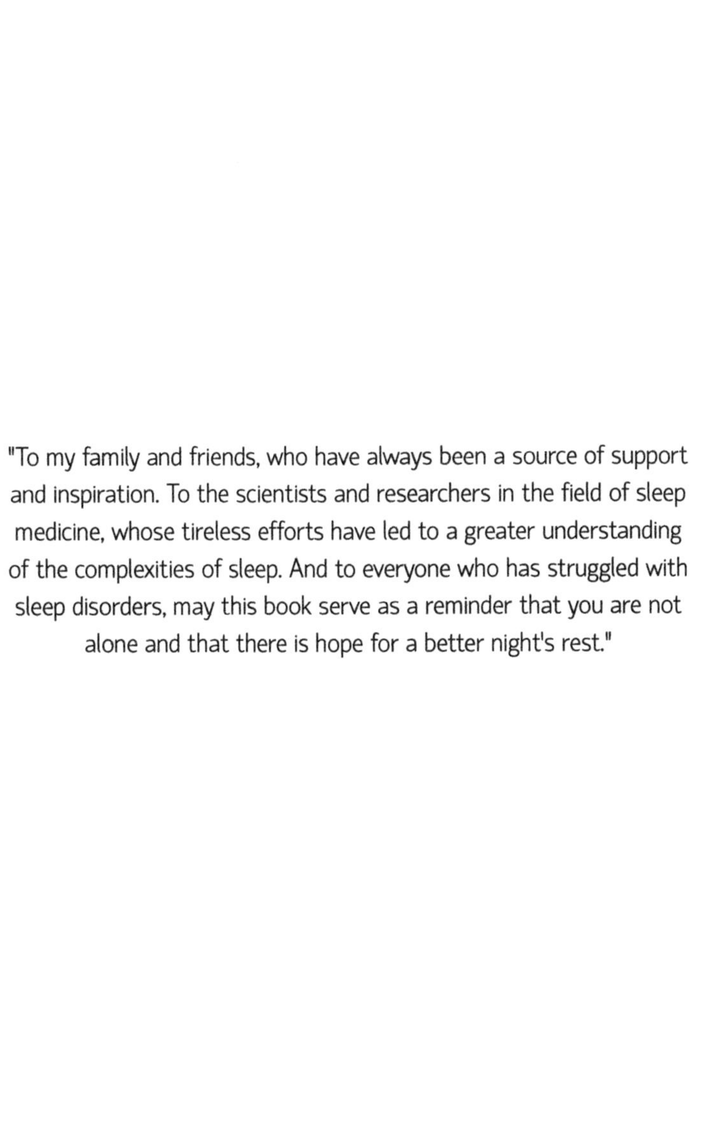

"To my family and friends, who have always been a source of support and inspiration. To the scientists and researchers in the field of sleep medicine, whose tireless efforts have led to a greater understanding of the complexities of sleep. And to everyone who has struggled with sleep disorders, may this book serve as a reminder that you are not alone and that there is hope for a better night's rest."

Contents

Foreword	*vii*
Preface	*ix*
Acknowledgements	*xi*
Prologue	*xiii*
1. Introduction	1
2. The Sleep Cycle	4
3. The Importance Of Quality Sleep	8
4. Common Sleep Disorders	11
5. Improving Sleep Quality	13
6. Medications And Treatments	15
Conclusion	21
Summary For Key Takeaways	23
Tips For Maintaining Healthy Sleep Habits	25

Foreword

The science of sleep is a complex and fascinating field that has captivated researchers and scientists for decades. As we continue to learn more about the intricacies of sleep, it becomes increasingly clear that sleep plays a critical role in maintaining overall health and well-being.

This book, "The Science of Sleep," delves into the various aspects of sleep, including the stages of the sleep cycle, the importance of quality sleep, and the common sleep disorders that can affect our daily lives. It also explores ways to improve sleep quality, and medications and treatments that are available.

The author has done an excellent job of presenting the information in a clear and concise manner, making it accessible to a wide audience. Whether you are a healthcare professional, a scientist, or simply someone who is curious about sleep, this book is an invaluable resource that will provide you with a deeper understanding of the science of sleep.

As we continue to learn more about the complexities of sleep, it is my hope that this book will serve as a reminder of the importance of sleep and the impact it has on our lives. I strongly recommend "The Science of Sleep" to anyone who is interested in learning more about this fascinating subject.

Khushdil shah
WRITER

Preface

Preface:

Sleep is an essential aspect of our lives that plays a critical role in maintaining overall health and well-being. However, despite its importance, many people struggle with sleep disorders that can have a significant impact on their daily lives.

This book, "The Science of Sleep," aims to provide a comprehensive overview of the various aspects of sleep, including the stages of the sleep cycle, the importance of quality sleep, and the common sleep disorders that can affect our daily lives. It also explores ways to improve sleep quality, and medications and treatments that are available.

As a researcher and healthcare professional, I have had the opportunity to work with individuals who struggle with sleep disorders. Through this experience, I have come to realize the importance of understanding the science of sleep, not only for those who suffer from sleep disorders but also for the general population.

In writing this book, my goal was to provide a comprehensive and accessible resource that would be beneficial to both healthcare professionals and the general public. I hope that this book will serve as a valuable resource for anyone who is interested in learning more about the science of sleep and the impact it has on our lives.

KHUSHDIL SHAH

WRITER

Acknowledgements

Writing a book is a journey that is never undertaken alone. There are many people who have helped and supported me along the way, and I am deeply grateful to each and every one of them.

First and foremost, I would like to thank my family and friends for their unwavering support and encouragement throughout the writing process. Their understanding and patience has been invaluable.

I would also like to extend my heartfelt thanks to the experts and professionals in the field of sleep science who have shared their knowledge and insights. Their contributions have been invaluable in ensuring that the information in this book is accurate and up-to-date.

I am also grateful to my editor and publisher for their support and guidance. Without their help, this book would not have been possible.

Finally, I would like to thank all of the individuals who have shared their stories and experiences with me. Their insights have been invaluable in helping me to understand the impact that sleep disorders can have on people's lives.

I hope that this book will be a valuable resource for anyone who is interested in learning more about the science of sleep and the impact it has on our lives.

Prologue

In this book, we will delve into the fascinating world of sleep and explore the latest scientific research on the subject. We will take a closer look at the different stages of sleep, the importance of sleep for our physical and mental well-being, and the potential consequences of sleep deprivation.

We will discuss the role that sleep plays in regulating our hormones, maintaining our immune system, and supporting our overall health. We will also explore the impact that sleep has on our cognitive abilities, including memory, attention, and learning.

We will also take a closer look at the various sleep disorders that can affect our sleep, such as insomnia, sleep apnea, and restless leg syndrome. We will discuss the causes and symptoms of these disorders, as well as the available treatment options.

In this book, we will also examine the connection between sleep and other areas of our lives, such as work, exercise, and diet. We will explore the latest research on the impact of technology and artificial light on our sleep, and discuss practical strategies for improving our sleep habits.

Through a combination of scientific research, case studies, and expert advice, this book will provide readers with a comprehensive understanding of the science of sleep and the importance of getting a good night's rest. Whether you're struggling with sleep issues or just looking for ways to improve your sleep, this book will provide you with the information and tools you need to achieve your best sleep yet.

CHAPTER ONE

INTRODUCTION

Sleep is an essential aspect of our lives that plays a critical role in maintaining overall health and well-being. It is during sleep that our bodies and minds are able to rest, repair, and rejuvenate. Adequate sleep is necessary for physical, mental and emotional health, as well as for maintaining cognitive function, memory, and learning.

However, despite its importance, many people struggle with sleep disorders that can have a significant impact on their daily lives. Insomnia, sleep apnea, restless leg syndrome, and narcolepsy are just a few examples of the many sleep disorders that can affect individuals. These disorders can lead to feelings of fatigue, irritability, difficulty focusing, and can even increase the risk of serious health conditions such as heart disease, diabetes, and stroke.

It is clear that adequate and quality sleep is necessary for overall health and well-being. This book, "The Science of Sleep," aims to provide a comprehensive overview of the various aspects of sleep, including the stages of the sleep cycle, the importance of quality sleep, and the common sleep disorders that can affect our daily lives. It also explores ways to improve sleep quality, and medications and treatments that are available.

The goal of this book is to provide a valuable resource for anyone who is interested in learning more about the science of sleep and the impact it has on our lives. Whether you are a healthcare professional, a scientist, or simply someone who is curious about sleep, this book is an invaluable resource that will provide you with a deeper understanding of the science of sleep.

Sleep disorders are a common problem that affects many individuals and can have a significant impact on overall health and well-being. Sleep disorders are conditions that affect the ability to fall asleep, stay asleep, or get restful sleep. Some common sleep disorders include insomnia, sleep apnea, restless leg syndrome, and narcolepsy.

Insomnia is a disorder characterized by difficulty falling asleep or staying asleep, leading to feelings of fatigue and irritability during the day. Sleep apnea is a disorder characterized by pauses in breathing during sleep and can lead to a lack of oxygen to the brain, which can have serious health consequences. Restless leg syndrome is a disorder characterized by an overwhelming urge to move the legs and can cause discomfort and difficulty falling asleep. Narcolepsy is a disorder characterized by excessive daytime sleepiness and can cause sudden sleep attacks during the day.

Each of these disorders has specific symptoms and causes, and can have a wide range of negative effects on the body and mind. Understanding the causes, symptoms, and effects of these common sleep disorders is crucial in order to address and manage them effectively. This book, "The Science of Sleep," delves into the various aspects of sleep disorders, including their causes, symptoms, and effects, as well as available treatments and ways to improve sleep quality.

Whether you are a healthcare professional, a scientist, or simply someone who is curious about sleep disorders, this book is an invaluable resource that will provide you with a deeper understanding of the science of sleep disorders and the impact they have on our lives.

CHAPTER TWO

The Sleep Cycle

Sleep is a natural state that is characterized by a decreased level of consciousness and a reduced ability to respond to the environment. Sleep is divided into two main states: non-rapid eye movement (NREM) sleep and rapid eye movement (REM) sleep.

NREM sleep is further divided into three stages: Stage 1, Stage 2, and Stage 3. Stage 1 is the lightest stage of sleep, characterized by a decreased level of consciousness, and a reduced ability to respond to the environment. During this stage, the brain waves are slow and irregular, and muscle activity is minimal. Stage 2 is a deeper stage of sleep, characterized by a further decrease in consciousness and a reduction in muscle activity. Brain waves during this stage are slower and more regular than during Stage 1. Stage 3, also known as slow-wave sleep, is the deepest stage of sleep, characterized by the slowest brain waves and the least muscle activity.

REM sleep is characterized by an increase in brain activity and muscle tone, as well as a decrease in body temperature and heart rate. This stage is associated with vivid dreams, as well as a temporary paralysis of the muscles, known as REM atonia.

During a normal night's sleep, an individual will cycle through these stages multiple times. It typically takes about 90 minutes to move through the entire cycle, and each stage lasts for a different amount of time. The first cycle of the night will be shorter and will mostly consist of NREM sleep, while later in the night, the cycles will be longer and will contain more REM sleep.

It's essential to note that the duration and timing of these stages can vary from person to person and may be affected by factors such as age, sleep disorders, and certain medications.

The sleep cycle is regulated by a complex interplay of hormones and brain activity. Hormones such as melatonin, which is produced by the pineal gland, and cortisol, which is produced by the adrenal glands, play a significant role in regulating the sleep-wake cycle.

Melatonin, also known as the "sleep hormone," is responsible for regulating the body's internal clock and signaling the brain that it is time to sleep. Melatonin levels increase in the evening as darkness falls, promoting feelings of sleepiness, and decrease in the morning as light increases, promoting feelings of alertness.

Cortisol, also known as the "stress hormone," plays a role in regulating the sleep-wake cycle by increasing alertness and energy levels during the day. The levels of cortisol are at their highest in the morning and decrease throughout the day, reaching their lowest levels at night, promoting feelings of sleepiness.

The brain also plays a critical role in regulating the sleep cycle through the activity of certain nerve cells and neurotransmitters. The hypothalamus, a region of the brain, contains nerve cells that control the sleep-wake cycle. These nerve cells are activated by neurotransmitters such as adenosine, which builds up during the day and promotes feelings of sleepiness, and dopamine, which promotes feelings of alertness.

Additionally, the activity of certain brain regions such as the reticular activating system, the hypothalamus and the brainstem are also involved in the regulation of the sleep-wake cycle.

Overall, it's clear that the sleep cycle is regulated by a complex interplay of hormones and brain activity, with a delicate balance between different hormones, neurotransmitters, and brain regions. Any disruption to this balance can lead to sleep disorders and can affect the quality of sleep.

Disruptions to the sleep cycle can have a wide range of negative effects on the body and mind. Some of the most common effects of disruptions to the sleep cycle include:

Fatigue and daytime sleepiness: When the sleep cycle is disrupted, individuals may have difficulty falling asleep, staying asleep, or getting restful sleep. This can lead to feelings of fatigue and daytime sleepiness, making it difficult to focus and perform daily activities.

Irritability and mood swings: Sleep deprivation can also affect emotional regulation, leading to irritability and mood swings.

Cognitive impairment: Disruptions to the sleep cycle can also affect cognitive function, memory, and learning. Studies have shown that individuals who are sleep-deprived have difficulty with attention, concentration, problem-solving, and decision-making.

Increased risk of accidents: Fatigue and sleepiness can also increase the risk of accidents, particularly when operating machinery or driving.

Increased risk of chronic health conditions: Chronic sleep deprivation can also increase the risk of serious health conditions such as heart disease, diabetes, and stroke.

Increased risk of mental health disorders: Disruptions to the sleep cycle can also increase the risk of mental health disorders such as depression and anxiety.

It's important to note that disruptions to the sleep cycle can be caused by a wide range of factors, including sleep disorders, lifestyle habits, and certain medications. By understanding the causes and effects of disruptions to the sleep cycle, individuals can take steps to address and manage them effectively.

CHAPTER THREE

The Importance of Quality Sleep

The quantity and quality of sleep are critical factors in maintaining overall physical and mental health.

Adequate quantity of sleep is necessary for physical health, allowing the body to repair and rejuvenate itself. Studies have shown that individuals who get enough sleep have a stronger immune system, which can help prevent illness and disease. Adequate sleep is also essential for maintaining a healthy weight, as it helps to regulate the hormones that control hunger and metabolism.

Quality of sleep is also important for physical health, as poor sleep quality can lead to a range of health problems. Poor sleep quality is associated with an increased risk of chronic health conditions such as heart disease, diabetes, and stroke. It's also associated with an increased risk of obesity, as well as a greater likelihood of accidents.

In terms of mental health, adequate sleep quantity and quality are essential for maintaining emotional regulation and cognitive function. Sleep deprivation can lead to irritability, mood swings, and difficulty with attention, concentration, problem-solving, and decision-making. Chronic lack of sleep is also associated with an increased risk of depression and anxiety.

Overall, it's clear that the quantity and quality of sleep are critical factors in maintaining overall physical and mental health. By understanding the importance of sleep and implementing healthy sleep habits, individuals can take the necessary steps to ensure that they are getting the rest they need.

The relationship between sleep and memory is a complex one, with research showing that sleep plays a crucial role in the consolidation and stabilization of memories. During sleep, the brain processes and organizes information from the day, strengthening connections between neurons and solidifying memories. Studies have shown that people who get a good night's sleep after learning something new perform better on memory tests compared to those who stay awake.

One of the stages of sleep, known as slow-wave sleep, is particularly important for memory consolidation. During this stage, the brain experiences slow oscillations in electrical activity, which are thought to be associated with the strengthening of connections between neurons involved in forming memories. The stage of sleep called Rapid Eye Movement (REM) sleep is also thought to play a role in memory consolidation, particularly for emotional memories.

Research has also shown that sleep deprivation can have a negative impact on memory and learning. Studies have found that people who are sleep-deprived have a harder time retaining new information and have a harder time with tasks that require attention, focus and decision making. Additionally, sleep deprivation has been linked to a decline in cognitive abilities, such as memory and learning.

In addition to the direct effects of sleep on memory, sleep also plays an indirect role in memory by supporting overall brain health. Adequate sleep is necessary for maintaining the overall health of the brain, including the growth of new neurons and the repair of damaged ones. This is important for maintaining cognitive function and preventing age-related decline in memory and learning.

In conclusion, sleep plays a crucial role in the consolidation and stabilization of memories, and the lack of it can have a negative impact on memory and learning. Regular, good quality sleep is essential for maintaining cognitive abilities and overall brain health.

Chronic sleep deprivation can have a number of negative effects on the body and mind. Physically, it can lead to an increased risk of obesity, diabetes, cardiovascular disease, and even cancer. It can also weaken the immune system, making it more difficult to fight off infections. Mentally, chronic sleep deprivation can lead to problems with memory and concentration, as well as an increased risk of depression and anxiety. It can also lead to irritability, mood swings, and impaired judgment. In addition to these health effects, chronic sleep deprivation can also lead to poor performance at work or school, and an increased risk of accidents or injuries. Overall, getting enough sleep is essential for maintaining good health and well-being.

CHAPTER FOUR

Common Sleep Disorders

Insomnia: Insomnia is a sleep disorder characterized by difficulty falling asleep or staying asleep. Causes of insomnia can include stress, anxiety, depression, certain medications, and certain medical conditions. Symptoms of insomnia include difficulty falling asleep, difficulty staying asleep, waking up frequently during the night, and feeling tired upon waking. Treatment options for insomnia include cognitive behavioral therapy, relaxation techniques, and medications such as sedatives and antidepressants.

Sleep Apnea: Sleep apnea is a sleep disorder characterized by pauses in breathing or shallow breathing during sleep. Causes of sleep apnea can include obesity, smoking, alcohol consumption, and certain medical conditions. Symptoms of sleep apnea include loud snoring, pauses in breathing during sleep, and feeling tired upon waking. Treatment options for sleep apnea include the use of a continuous positive airway pressure (CPAP) machine, lifestyle changes such as weight loss and avoiding alcohol, and surgery in some cases

Restless Leg Syndrome: Restless Leg Syndrome (RLS) is a sleep disorder characterized by an overwhelming urge to move the legs, usually accompanied by uncomfortable sensations such as tingling, burning, or aching. Causes of RLS can include genetics, iron deficiency, and certain medical conditions. Symptoms of RLS typically occur in the evening and night, and can make it difficult to fall asleep and stay asleep. Treatment options for RLS include medications such as dopaminergic drugs and iron supplements, as well as lifestyle changes such as regular exercise and avoiding caffeine and alcohol.

Narcolepsy: Narcolepsy is a sleep disorder characterized by excessive daytime sleepiness and sudden sleep attacks. Causes of narcolepsy can include genetics and certain medical conditions. Symptoms of narcolepsy include excessive daytime sleepiness, sudden sleep attacks, vivid hallucinations, and temporary paralysis upon waking. Treatment options for narcolepsy include medications such as stimulants and antidepressants, as well as lifestyle changes such as maintaining a consistent sleep schedule

CHAPTER FIVE

Improving Sleep Quality

Improving sleep quality is essential for maintaining overall health and well-being. Some ways to improve sleep quality include:

Maintaining a consistent sleep schedule

Creating a comfortable sleep environment (e.g. keeping the bedroom dark, quiet, and at a comfortable temperature)

Avoiding electronic devices for at least 30 minutes before bedtime

Engaging in relaxation techniques such as yoga or meditation

Avoiding caffeine and alcohol close to bedtime

Practicing good sleep hygiene.

Tips for creating a sleep-conducive environment:

Keep your bedroom dark, cool, and quiet

Use heavy curtains or blackout shades to block light from windows

Use earplugs or a white noise machine to block out noise

Use a comfortable mattress and pillows

Keep electronic devices such as phones and laptops out of the bedroom

Avoid bright screens for at least 30 minutes before bedtime

Sleep hygiene: best practices for a healthy sleep routine:

Stick to a consistent sleep schedule

Avoid caffeine and alcohol close to bedtime

Avoid heavy meals close to bedtime

Engage in regular physical activity during the day

Avoid electronic devices for at least 30 minutes before bedtime

Create a relaxing bedtime routine

Relaxation techniques for falling asleep:

Deep breathing exercises

Progressive muscle relaxation

Guided imagery

Yoga or stretching

Meditation

Listening to calming music

Alternative therapies for improving sleep:

Acupuncture

Herbal supplements such as valerian root or melatonin

Aromatherapy with essential oils such as lavender or chamomile

Massage therapy

Chiropractic care

Cognitive behavioral therapy

It is important to note that while these alternative therapies can be effective in improving sleep, they may not work for everyone, and it is always best to consult with a healthcare professional before starting any new treatment or supplement regimen.

CHAPTER SIX

Medications and Treatments

There are several medications and treatments available for sleep disorders. Some of the most commonly used include:

Sedatives: These medications, also known as hypnotics, are used to help people fall asleep. Examples include zolpidem (Ambien) and eszopiclone (Lunesta). Sedatives can be effective in the short-term, but they can also cause side effects such as grogginess, dizziness, and confusion. They may also lead to dependence and tolerance over time.

Antidepressants: Some antidepressants, such as trazodone and mirtazapine, have sedative properties and may be used to treat insomnia. They can also be used to treat other sleep disorders such as sleep apnea and RLS. These medications may have more side effects than sedatives, such as dry mouth, constipation, and weight gain.

Stimulants: Medications such as modafinil (Provigil) and armodafinil (Nuvigil) are used to treat narcolepsy and other sleep disorders characterized by excessive daytime sleepiness. These medications can cause side effects such as anxiety, headache, and nausea.

Continuous positive airway pressure (CPAP) machine: This is a device used to treat sleep apnea. It delivers a steady stream of air pressure through a mask worn over the nose or mouth, which helps to keep the airway open during sleep.

Lifestyle changes: In some cases, changes in lifestyle, such as weight loss, avoiding alcohol, and regular exercise, can help to improve sleep.

It is important to note that the selection of medication and treatment option depends on the specific sleep disorder and the individual's circumstances, and it is always best to consult with a healthcare professional before starting any new treatment or medication regimen.

The pros and cons of medications and treatments for sleep disorders can vary depending on the specific medication or treatment, as well as the individual's unique circumstances. Here is a general overview of some of the pros and cons of commonly used medications and treatments:

Sedatives: Pros:

Can help people fall asleep quickly

Often effective in the short-term

Cons:

Can cause side effects such as grogginess, dizziness, and confusion

Can lead to dependence and tolerance over time

Risk of accidents such as falls and car crashes due to the sedative effects

Antidepressants: Pros:

Can help with insomnia, sleep apnea and RLS

Can also help with other conditions such as depression and anxiety

Cons:

May have more side effects than sedatives, such as dry mouth, constipation, and weight gain

Can take several weeks to start working

Risk of interactions with other medications

Stimulants: Pros:

Can effectively treat narcolepsy and other sleep disorders characterized by excessive daytime sleepiness

Cons:

Can cause side effects such as anxiety, headache, and nausea

May not be recommended for people with certain medical conditions such as hypertension, heart disease, and mental health conditions

Risk of addiction and tolerance over time

Continuous positive airway pressure (CPAP) machine: Pros:

Can effectively treat sleep apnea

Has few side effects

Cons:

Can be uncomfortable to wear

Can be loud

Need to be cleaned regularly

May make it difficult to sleep if the mask is not comfortable

Lifestyle changes: Pros:

Can be effective in improving sleep

No side effects

Cons:

Can be difficult to implement and maintain

May not work for everyone

Can take time to show results

It is important to note that these are general pros and cons and may not apply to everyone. It is always best to consult with a healthcare professional to determine the best treatment option for the individual's specific sleep disorder and circumstances.

All medications and treatments for sleep disorders come with the potential for risks and side effects. These can vary depending on the specific medication or treatment, as well as the individual's unique circumstances. Here is a general overview of some of the risks and potential side effects of commonly used medications and treatments:

Sedatives:

Risk of accidents such as falls and car crashes due to the sedative effects

Dependence and tolerance can occur over time

Risk of withdrawal symptoms if the medication is stopped suddenly

Long-term use may lead to decreased effectiveness, rebound insomnia, and other complications

Antidepressants:

Risk of interactions with other medications

Risk of withdrawal symptoms if the medication is stopped suddenly

Long-term use may lead to decreased effectiveness, rebound insomnia, and other complications

Stimulants:

Risk of addiction and tolerance over time

May not be recommended for people with certain medical conditions such as hypertension, heart disease, and mental health conditions

May cause side effects such as anxiety, headache, and nausea

Continuous positive airway pressure (CPAP) machine:

May make it difficult to sleep if the mask is not comfortable

Need to be cleaned regularly

Risk of sinus infections or skin irritation from the mask

Lifestyle changes:

Can be difficult to implement and maintain

May not work for everyone

Can take time to show results

It is important to note that these are general risks and potential side effects and may not apply to everyone. It is always best to consult with a healthcare professional to determine the best treatment option for the individual's specific sleep disorder and circumstances, and to monitor the treatment progress regularly.

Conclusion

In conclusion, sleep is a vital aspect of our overall health and well-being. A variety of sleep disorders, such as insomnia, sleep apnea, restless leg syndrome, and narcolepsy, can disrupt our ability to get a good night's sleep. Understanding the causes, symptoms, and treatment options for these sleep disorders is essential for managing them effectively.

Medications and treatments such as sedatives, antidepressants, stimulants, continuous positive airway pressure (CPAP) machine and lifestyle changes can be used to improve sleep. However, it is important to consider the potential risks and side effects of these treatments, as well as the individual's unique circumstances.

Creating a sleep-conducive environment, practicing good sleep hygiene, and engaging in relaxation techniques can also help to improve sleep. Alternative therapies such as acupuncture, herbal supplements, aromatherapy, massage therapy, chiropractic care, and cognitive behavioral therapy may also be effective for some people.

It is important to remember that sleep disorders can be complex and multi-faceted, and the best treatment approach will depend on the individual's specific needs and circumstances. Consulting with a healthcare professional is the best way to determine the best course of action for managing sleep disorders and improving sleep quality.

Overall, this book has aimed to provide an understanding of the science of sleep and the various sleep disorders that can affect us. By learning about sleep and how it works, we can take steps to improve our sleep and lead healthier, happier lives.

Advancements in technology and research methods are allowing for a deeper understanding of the underlying mechanisms of sleep and sleep disorders. This can lead to the development of more effective treatments and therapies.

Research can help to identify new sleep disorders and better understand existing ones. This can lead to earlier diagnosis and more targeted treatments.

Ongoing research can also help to uncover the links between sleep and other health conditions, such as diabetes, heart disease, and mental health conditions.

Research can also help to identify risk factors for sleep disorders and develop interventions to prevent them.

With the increasing awareness of the importance of sleep, research is also needed to evaluate the effectiveness and safety of popular sleep aids and remedies.

Research can also help to understand the impact of lifestyle, environment, and other factors on sleep patterns and quality.

Ongoing research in sleep science is essential for the development of evidence-based guidelines for sleep medicine and for the education of healthcare professionals and the general public about sleep health.

In summary, ongoing research in sleep science is crucial for improving our understanding of sleep and sleep disorders and for developing more effective treatments and interventions to promote sleep health.

Summary For Key Takeaways

Some key takeaways from this book include:

Sleep is essential for maintaining good health and well-being.

A variety of sleep disorders, such as insomnia, sleep apnea, restless leg syndrome, and narcolepsy, can disrupt our ability to get a good night's sleep.

Understanding the causes, symptoms, and treatment options for these sleep disorders is essential for managing them effectively.

Medications and treatments such as sedatives, antidepressants, stimulants, CPAP machines, and lifestyle changes can be used to improve sleep.

Creating a sleep-conducive environment, practicing good sleep hygiene, and engaging in relaxation techniques can also help to improve sleep.

Alternative therapies such as acupuncture, herbal supplements, aromatherapy, massage therapy, chiropractic care, and cognitive behavioral therapy may also be effective for some people.

Consulting with a healthcare professional is the best way to determine the best course of action for managing sleep disorders and improving sleep quality.

To improve sleep and lead healthier, happier lives, we should be aware of our sleep patterns, needs, and seek help if we are facing any sleep disorder.

Tips For Maintaining Healthy Sleep Habits

Stick to a consistent sleep schedule by going to bed and waking up at the same time each day, even on weekends.

Create a comfortable sleep environment by keeping your bedroom dark, cool, and quiet, and using a comfortable mattress and pillows.

Avoid electronic devices for at least 30 minutes before bedtime, as the blue light emitted from screens can interfere with the production of melatonin, a hormone that helps regulate sleep.

Engage in relaxation techniques such as deep breathing exercises, progressive muscle relaxation, guided imagery, yoga or stretching, meditation or listening to calming music before bed.

Avoid caffeine and alcohol close to bedtime as they can disrupt sleep patterns.

Avoid heavy meals close to bedtime, try to finish your last meal 2-3 hours before you go to bed.

Engage in regular physical activity during the day, as regular exercise can help improve sleep quality.

Practice good sleep hygiene by keeping your sleep environment clean, quiet and dark, and making sure your bed is comfortable.

If you're having trouble falling asleep, get out of bed and engage in a relaxing activity in dim light, such as reading a book or listening to music.

If you still have trouble falling asleep after trying these tips, consult with a healthcare professional to determine if there is an underlying sleep disorder or medical condition that needs to be addressed.

9 798889 513193

Printed by Libri Plureos GmbH in Hamburg,
Germany